LIVERPOOL HISTORICAL ESSAYS

NEWTON AND THE CONCEPT OF MASS-ENERGY

by

PETER ROWLANDS

Published for the
DEPARTMENT OF HISTORY
UNIVERSITY OF LIVERPOOL

LIVERPOOL UNIVERSITY PRESS 1990

First published 1990 by
Liverpool University Press
P.O. Box 147
Liverpool
L69 3BX

British Library Cataloguing in Publication Data
Data are available
ISBN

Printed by Tudor Print, Bootle, Merseyside

CONTENTS

Preface		i
1	Newton and the concept of energy	1
2	The origin of the rest energy term in relativity	3
3	Material and radiation gases	6
4	The abstract nature of Newton's reasoning	8
5	The Newtonian calculation of light bending	11
6	Newton's calculations on refraction	14
7	Capillary action and the mass of a particle of light	17
8	The corpuscular and quantum theories of light	19
9	Conclusion	25
Notes and references		28
Appendices		31

PREFACE

Isaac Newton's *Opticks*, as is well-known, ended with a series of 31 *Queries* containing some of the most remarkably prescient guesses ever made by a practising scientist. Taken at face value, they were the last speculative efforts of an ageing natural philosopher on subjects which were not yet amenable to the more rigorous methods previously employed in his optics and dynamics. But Newton's optical Queries were not founded on speculation. They were founded on extended and precise chains of physical and mathematical reasoning, and were all that he dared to publish from a vast amount of manuscript material accumulated over a long period of years. The meaning of the Queries cannot be found in the published texts alone but must be sought also in the draft material from which they were composed.

One of the most interesting statements in all the Queries is that of the possible interconvertibility of light and matter in Query 30. Historians of science have often compared this statement with the modern doctrine of the interconvertibility of energy and mass summarised in Einstein's famous equation $E = mc^2$, but without any serious belief that the two theories could be *mathematically* connected. However, when Newton's drafts are more closely examined, a clear mathematical connection does emerge, and something even more remarkable still: *Newton's arguments help us to solve an anomaly in the modern theory*, the fact that relativity does not derive the relationship between mass and energy, but only the relationship between *changes* in the values of these quantities.

The direct relationship between the total mass of a body and its energy content, proposed by Einstein on inductive grounds, is undoubtedly true, but it cannot be derived from relativity theory; the work of Newton and some eighteenth and nineteenth century scientists, though containing no such explicit relation, tells us why. It also explains the curious fact that photon energy is mc^2, rather than $mc^2/2$. Newton, like his immediate successors, worked from an analogy between optics and steady state dynamics in which mc^2 was a *potential*, rather than kinetic, energy term. The analogy produced results meaningful in modern terms because of the validity of a general dynamic theorem known to him only in particular cases.

In unravelling Newton's optical drafts, we uncover a whole host of relationships between apparently unconnected branches of physics and between ideas developed by different scientists at different periods. Most importantly, perhaps, we learn that the fundamental meaning of a scientific truth is not necessarily to be found in the context in which it was first discovered.

Peter Rowlands
Oliver Lodge Laboratory
University of Liverpool

NEWTON AND THE CONCEPT OF MASS-ENERGY

1. Newton and the concept of energy

In a famous passage from Query 30 of the *Opticks*, Newton asked: "Are not gross Bodies and Light convertible into one another, and may not Bodies receive much of their Activity from the Particles of Light which enter their Composition?", while later in the same Query he made the definite assertion that: "The changing of Bodies into Light, and Light into Bodies, is very conformable to the Course of Nature, which seems to be delighted with Transmutations."[1] This has been seen as a prescient observation, an anticipation of the twentieth century view of the interconversion of energy and matter; but, associated as it is in the Query with such transformations as that of water into earth, it seems rather remote from the result

$$E = mc^2$$

derived from more precisely mathematical modern approaches. Newton's own approach to this question, however, was itself mathematical in origin, as can be seen from the material from which the Query was composed, and it was only because the bulk of this material was suppressed when the Query was finally published that Newton's statement seems to be nothing more than a brilliant speculation.

It is generally held that the equation $E = mc^2$ is a specific result of the special theory of relativity, that it cannot be derived from 'classical' approaches and that it is somehow bound up with the intrinsically 4-vector nature of relativistic space-time. It is thought inconceivable that Newton could have had any such idea in mind when he wrote Query 30; and, in any case, a classical calculation of kinetic energy, even for the case of solid light particles of mass m and velocity c, would still yield a value only half of that which mysteriously appears as a result of relativistic calculations. But the drafts for Query 22 of the Latin edition of the *Opticks* (1706), which ultimately became Query 30 of the subsequent English edition (1717), are based very strongly on the idea that the refraction of light is due to a force mc^2/r acting over distance r, and it is only a short step from a force of this form, acting under steady-state conditions, to a *potential* energy of the form mc^2.

Now Newton, as we know, unlike Leibniz, had no word or expression equivalent to the modern term 'energy' or, indeed, to any form of it, but it is not difficult to find examples of *calculation* in his work which are expressed unmistakably in energy terms. In the *Principia*, for example, he clearly used the quantities which we now describe as work, kinetic energy, potential energy and power. In Book I, Proposition XXXIX, which considered the velocity of a rising or falling body produced by the action of an arbitrary force, he stated, according to R. S. Westfall, "the mechanical equivalent of the work-energy equation, expressed indeed in a more general form than Leibniz ever achieved"; in precise terms, the work done, the integral of force over distance, was equal to the change in kinetic energy ($\Delta W = \Delta(mv^2/2)$).[2] The third corollary to the proposition stated the same fact even more precisely, while the first corollary related the kinetic energy to the change in kinetic energy as $W/\Delta W = v^2/2v\Delta v$. Proposition XL showed that the sum of potential and kinetic energies in gravitational and other orbits remains constant with time,[3] while Proposition XLI, which D. T. Whiteside has described as the central dynamic theorem of the *Principia*, was a version of the well-known 'vis viva' integral, a conservation of energy equation used for finding the paths taken by bodies subject to any type of centripetal force.[4] Book II, Proposition XXX even extended the analysis to resisted motion; here it was shown that, in the case of a pendulum moving through a resisting medium, the "loss of potential energy due to resistance is equal to the work done against resistance".[5] Finally, Newton's analysis of machines in the Scholium to the Laws of Motion suggested that the rate of energy consumption, calculated as the product of force and velocity, would be equal to the sum of the useful power generated and the power dissipated by friction of the parts.[6]

Newton was thus well aware of the nature of the integral of force over distance and of its relationship with terms involving mass and velocity squared; and though he expressed it in Proposition XXXIX in terms of the area of a geometrical diagram, his use of the phrase "granting the quadrature of curvilinear figures" indicates that this particular theorem was completely analytic in origin and that the relationship between work and energy was derived by him in essentially the form we derive it today. In principle, he recognised that, for purposes of *calculation*, it was as valid to state equations using the mathematical construct which we now describe as 'energy' as it was to work from the laws of motion based on the concept which he had defined as 'force'. At a later date, Newton's close disciple, Samuel Clarke, working probably under his direct inspiration and certainly with his complete approval, showed a clear understanding of the Leibnizian concept of vis viva (mv^2) and how it related to the Newtonian concept of force.[7] Essentially, then, Newtonian relations involving a force term such as mc^2/r necessarily presupposed an equivalent energy equation involving terms of the form mc^2.

Perhaps even more remarkable is the fact that the term mc^2, or its verbal equivalent, actually occurs in Newton's early work in optics. The essay "Of Colours", which is thought to date from about 1666, contains a calculation of the enlargement to be expected in the area of the interference rings produced in a thin film when viewed through an oblique angle. Here, the "factus of ye motion of ye incident ray into its velocity",[8] which is the measure of the strength of the ray, is the product of the momentum ($p = mc$) and the perpendicular component of the velocity ($c \cos i$), and, when the ray is perpendicular, is precisely mc^2. The areas of the rings, in Newton's formula, were inversely proportional to the energies due to the components of their motions perpendicular to the film.

2. The origin of the rest energy term in relativity

The relationship

$$E = mc^2$$

is derived in special relativity by a non-deductive method, in which it is effectively brought in as an assumption. It arises as an unexplained term in the integral of the Lorentz-invariant equation

$$\frac{dT}{dt} = \mathbf{F} . \mathbf{v}$$

relating the rate of increase of kinetic energy (T) to the rate at which work is done. From this equation we derive

$$T = \frac{mc^2}{(1 - v^2/c^2)^{1/2}}$$

which expands to give

$$T = mc^2 + \frac{mv^2}{2} + \dots \quad ,$$

Here the term mc^2 is a constant added to the classical $mv^2/2$ and can be considered arbitrary since we could add any constant of integration to the equation. Adding $-mc^2$, for example, would remove the term entirely and make the expression, for $v \ll c$, identical with the classical one, as would normally be required. It is usually found *convenient*, in relativity, to take the constant of integration as 0, because this allows a convenient definition of a 4-vector momentum, and then to find a 'physical' meaning for the added term mc^2. The rather arbitrary and non-deductive nature of this process suggests that, like so many other things in relativity (the Schwarzschild radius, the equations for the expanding universe, the gravitational redshift), the expression does not arise from the theory of relativity itself but is a more fundamental truth which that theory has uncovered.

Remarkably, it is the 'classical' work of Newton, on the bending of light by refraction, which may indicate what that more fundamental truth may be. The expression mc^2 for electromagnetic energy stems from nothing more than the fact that it is a potential energy term which, by the virial theorem, takes twice the normal value of kinetic energy in any steady state system subject *either* to constant or to inverse square law forces; for a constant force it is positive and, for an inverse square force, negative. The same relation between kinetic and potential energies emerges in all such steady-state systems regardless of the physical nature of the force or type of potential which is assumed to operate, and, if potential energy is replaced by the work done, it applies (in a more general form) to practically any system not subject to dissipative forces. The difference between the numerical value for the potential energy or total work done in such a system and the normal expression for the kinetic energy induced when work is done in changing the speed of a body is due solely to the steady state conditions which apply in the former instance, and it is notable that the *half*-value $mc^2/2$ is used as the energy term in cases where systems are described in specifically *kinetic* terms; for example, in the derivation of the Schwarzschild radius, where c is taken as the maximum speed of a particle escaping from a gravitating body.

The potential nature of mass-energy should be perhaps obvious from its incorporation of the idea of 'rest' energy; energy due to matter at rest can hardly be 'kinetic'. Even photon energy can be described as 'potential' in this sense, for the photon only has existence as a 'particle' at the point of interaction.

Again, the intrinsically arbitrary nature of the numerical value is typical of potential energy terms. Writers who have investigated Einstein's own arguments and who demonstrate the validity of his derivation of the equation

$$\Delta E = \Delta mc^2$$

point to the arbitrary, though physically reasonable, nature of its extension to a body's rest mass. Stachel and Toretti, for example, state that: "The final conclusion that the entire mass of a body is in effect a measure of its energy, is of course entirely unwarranted by Einstein's premises";[9] and they quote Einstein as follows: "A mass μ is equivalent – insofar as its inertia is concerned – to an energy content of magnitude μc^2. Since we can arbitrarily fix the zero of (the rest energy), we are not even able to distinguish, without arbitrariness, between a 'true' and an 'apparent' mass of the system. It appears much more natural to regard all material mass as a store of energy."[10] Of course, what is arbitrary in special relativity need not be arbitrary in other contexts; if an idea is 'physically reasonable' or 'natural', it must be explicable in terms of some definite physical principles; and if mass is to be considered as a 'store' of energy, then this principle must be related to the idea of mass as a specifically *potential* form of energy.

It is not difficult now to see that the redefinition of mc^2 as a classical potential term solves an anomaly in relativistic physics, for the transition from special relativistic to classical mechanics at low speeds does not follow unless mc^2 *also has a classical meaning*; and it also explains why the mass-energy equation does not arise as a deductive consequence of the special theory of relativity, but must be derived *inductively* from the special relativistic equation $\Delta E = \Delta mc^2$. In precise terms, $E = mc^2$ is a classical equation and its justification in relativistic physics is that it represents the classical limit to which relativistic mechanics converges under appropriate conditions.

3. Material and radiation gases

The fact that mc^2 is essentially an expression for *potential* energy is shown by the classical (Boltzmann) derivation of the radiation law from a 'radiation gas', entirely analagous to the gas of material particles assumed in classical kinetic theory. The potential energy equivalent for the material gas is derived from the product of pressure and volume, is twice the total kinetic energy of the molecules moving in any given direction and is $mc^2/3$ for each degree of freedom; similarly, the energy of the radiation gas is the product of the radiation pressure and volume and is also $mc^2/3$ for each degree of freedom. The expression mc^2 clearly does not depend in any way on the 4-vector or 'relativistic' properties of light, for the analagous system of gaseous molecules is not 4-vector.

Of course, it is usual in the kinetic theory of material gases, to assume that the pressure produced by the gas is purely a result of collisions between the molecules and the container walls, but this is merely an operational convenience; pressure terms of any kind, whatever their origin, are an expression of the action of force or potential energy. A gas in steady state is equivalent to a system with constant expansive force in all directions, and a system of this kind necessarily requires a virial relation of the form

$$\overline{V} = 2\,\overline{T}$$

between the time-averaged potential and kinetic energies.

Newton's use of the term mc^2/r as the force exerted on light particles during refraction is entirely consonant with the use of mc^2 as the potential energy of the system, in the same way as mv^2 is the potential energy for bodies kept in gravitational orbit by the centripetal force mv^2/r. This is because the constancy of the velocity of light ensures that the optical system is 'steady-state' and that its potential energy term is numerically equivalent to that in the inverse-square-law gravitational system. The gravitational case is a particular example of the virial theorem with which Newton would certainly have been familiar. Another is his formula for the velocity of waves in a medium, in terms of elasticity or pressure (E) and density (ρ), which he applied to both light and sound, Essentially, Newton's formula

$$c = (E/\rho)^{1/2}$$

is an expression of the fact that the potential energy of the system of photons or gas molecules (mc^2) is equal to the work done at constant pressure as product of pressure and volume.

An application to light occurs in the calculation in Query 21 of the ratio between the elasticities per unit density of Newton's proposed electro-optic 'aether' and atmospheric air,[11] and the manuscript evidence shows that this was linked also with the calculation of the force of refraction. Newton's elasticity of the aether is what we would call energy density of radiation (ρc^2), which is related by Maxwell's classical formula of 1873 to the radiation pressure, and the ratio he calculates is, in effect, the ratio of the energy per unit mass of a particle of light to the energy per unit mass of an air molecule, as manifested in the transmission of sound. Now, the molecular potential energy per unit mass for air may be calculated from the kinetic theory of gases ($PV = RT$) at about 1.8×10^5 J kg^{-1} when T = 300 K. Since the energy per unit mass for a light photon is 9×10^{16} J kg^{-1}, the ratio is 5×10^{11}, which is comparable with Newton's 4.9×10^{11} minimum in Query 21.[12] The ratio has the particular significance of being the potential energy per unit mass ratio characteristic of the two highest levels of organization in matter, the interatomic and the intermolecular, and it relates to Newton's observations elsewhere that a higher level of force or energy is required to expel light particles in chemical reactions than to expel molecules from liquids as gases and vapours.

Newton's formula for calculating the velocities of waves in a medium is a perfect illustration of an application of the virial theorem, and it is because it is such a perfect illustration that it works in the case of light waves travelling in a vacuum, where the model of interaction with matter no longer applies. The centripetal force calculation used in Newton's studies of refraction is yet another such illustration, and must give the correct numerical energy relation whether or not the description of the force is correct. The constraints which need to be applied to find the true nature of the vector force are not required to find the numerical value of the scalar energy.

It is clear that Newton knew exactly what he was doing in these calculations. Several of his theorems in the *Principia* show how the same energy equations apply to many different types of force condition. Light, of course, always gives the correct results when travelling through a vacuum, because in such circumstances there is no source of dissipation, and the virial relation takes on its ideal form. This is why the elasticity of light is precisely the same thing as its energy density. Sound in air, as Newton realised, does not give perfect results, because, as we now know (though he did not), the relatively slow compression and rarefaction of the gas prevents the system from maintaining thermal equilibrium. Laplace, of course, subsequently showed that agreement between theory and experiment is reached, even in this case, if the elasticity is taken as adiabatic rather than isothermal.

4. The abstract nature of Newton's reasoning

All the results so far considered show how Newton's preferred method of theoretical investigation was in terms of general abstract concepts rather than of concrete or mechanistic models. His analysis of Boyle's law is a good example.[13] Newton derives the fact that pressure in a gas is proportional to density from an assumption that there is a force between adjacent gas molecules inversely proportional to the distance between them, but he makes a point of denying any belief in the validity of a *physical* model of a gas composed of stationary molecules subject to mutually repulsive forces. In fact, the derivation of Boyle's law from the modern kinetic theory of gas molecules in random motion is essentially the same as that of Newton from the static model, for the kinetic behaviour of the molecules is not a factor in deriving the potential energy term with which Boyle's law is really concerned. It is a purely formal matter whether we describe the gas in terms of the average kinetic energy of the individual molecules or an equivalent averaged-out potential energy of the gas as a whole.

In the modern theory, we assume in effect that the gas molecules are stationary overall – that is, that they occupy the same positions on a time average – and calculate the outward force exerted by each molecule to be inversely proportional to the length of the container in which it is placed or to any fixed portion of this length; this is exactly the same thing as making the outward force exerted by any molecule on its nearest neighbour inversely proportional to the separation of the two molecules. (It is not, of course, necessary to assume that this is due to a physical *interaction* between the molecules.) It is only by applying the virial theorem to the potential energy term PV, derived from Boyle's law, that we are able to bring in the kinetic nature of the gas and to explain its temperature behaviour.

Newton's very abstract approach to this problem may be illustrated by the fact that he derived, for the general case of forces inversely proportional to the power n of the distance, the corresponding proportionality of pressure to the power (n + 2)/3 of the density. It is by the same kind of argument that, in the virial theorem, we show that, for forces inversely proportional to power n of distance or for potential energies inversely proportional to power (n – 1), the time-averaged kinetic and potential energies are related by the formula

$$\overline{T} = \frac{(1 - n)}{2}\overline{V} \quad .$$

Newton, like Maxwell after him, used mechanistic models to help develop his mathematical insights, but he was well aware that the fundamental truths were intrinsically abstract and were independent of the model he had used to develop them.

It is likely, in fact, that the expression $E = mc^2$ and the ideal form of the virial theorem are truths which are more fundamental than dynamics itself, on the same level as the conservation of mass and charge, or the relationship between such conservation laws and fundamental symmetries. They are archetypal truths whose validity is independent of the context in which they were first discovered and are no more restricted to the terms of some particular physical theory than the so-called 'Second Law of Thermodynamics' is restricted to the terms of thermodynamics; we may, therefore, expect to find them in 'classical', as well as in relativistic, contexts. Special relativity may point to the universality of the mass-energy relation and extend it from the particular case of the light quantum to all other types of matter but it is not the ultimate source of the relation and cannot be used to derive it.

The power of the virial theorem is shown by the way in which it served, if somewhat unconsciously, to link several of Newton's separate investigations and allowed him to develop unexpected mathematical analogies between apparently unrelated phenomena. In Newton's physics the centripetal force law mv^2/r is closely linked with systems acting under the force of gravity, and it was by comparison with such systems that he developed his calculations on the force of refraction. It is significant that Newton's earliest derivation of the centripetal force law involved essentially the same collision process (involving aether particles) as is now used in the kinetic theory of gases; here, force is calculated as the product of the change in momentum in a particle due to impact and the rate at which collisions take place, the collision rate being found by dividing the particle velocity by the distance travelled between collisions.[14] Newton subsequently derived it, without using the impact mechanism, from the geometry of the circle alone, and he used similarly abstract arguments when he showed that an inverse square law of attraction was the only physically probable law which would explain the observed orbits of the planets.

However, although he treated gravity abstractly in the *Principia*, Newton was always prepared to consider the possibility that it could be explained by an impact mechanism involving aether particles. His early theories of gravity, to which he returned at intervals, involved a stream of aether particles of density ρ producing pressure ρv^2;[15] one of these theories, based on an idea of his follower, Fatio de Duillier, even involved a concept of aether particles moving randomly in straight lines until collisions caused changes of direction.[16] It was a standard notion, recurring throughout Newton's writings, that aether was, or acted like, a highly rarefied form of 'air', and so he must have had knowledge of the fact that, mathematically, dynamic models of both air and aether produced the same law of forces (mv^2/r) as those based on a central inverse square law force. The reason for this is the apparent coincidence that systems acting under inverse square law forces and steady state dynamic systems subject to effectively constant force produce the same numerical relation between potential energy and mv^2 or between force and mv^2/r;[17] but, coincidence or not, it was a fact of which Newton was aware after his early calculations of centripetal force, and he must have chosen the static model for his derivation of Boyle's law mainly on account of its simplicity.

The result, for Newton, was that his model for the force of refraction was mathematically equivalent to his formula for the velocity of light in aether, in the same way as the equation of motion of a satellite under gravity is mathematically equivalent, under the right conditions, to the formula for the velocity of sound in a medium.[18] With the further stipulation that particles in random motion may be treated as having independent velocities in each of three perpendicular directions, these become equivalent to the respective calculations for pressure in radiation and material gases, and, indeed, to Newton's calculations involving the pressures caused by optical and gravitational aethers. It is not surprising, therefore, that several of the early exponents of the kinetic theory of gases, for example, Le Sage, Herapath and Waterston, were inspired by the possibility that this theory might explain gravity.

5. The Newtonian calculation of light bending

Though it is certainly the comparison between the calculations for pressure and velocity in material and radiation gases which gives the clearest indication that the term mc^2 in quantum theory is exactly equivalent to the potential energy term mv^2 in classical physics, it is possible to find evidence also for the analogy in Newton's own preferred instance of centripetal forces acting on light rays, though this time the calculation is not due to Newton himself. The independence of $E = mc^2$ from the more general assumptions of 4-vector theory is demonstrated in a very striking way by the classical or 'Newtonian' derivation of the bending of a light ray.

Cavendish, in an unpublished calculation of about 1784, and Soldner, in a published calculation of 1801, predicted that a light ray passing at a distance of closest approach r from a heavy body of mass M would experience an angular deflection of $2GM/rc^2$. This was based on a typically Newtonian-type calculation for a hyperbolic orbit of eccentricity e in which

$$c^2 = GM\,(e - 1)/r$$

the hyperbolic, rather than elliptical or circular, orbit being a result of the exceptional speed of the light ray.[19] For a light particle of mass m, this is precisely the same result as would be obtained from a potential energy equal to mc^2.

It is, of course, well-known that this is only half the total deflection, and it is attributed in relativity specifically to the effect of time dilation or gravitational redshift; Einstein derived it in this way in 1911 using special relativity and the principle of equivalence,[20] and elsewhere he used the same assumptions in calculating the redshift. But it is, in fact, completely unnecessary to use either the principle of equivalence or special relativity to derive the gravitational redshift. We need only use the quantum theory and $E = mc^2$, and to first order we need only assume that a quantum of energy $h\nu$ and mass $h\nu/c^2$ is subjected to a gravitational potential $-GM/r$ to produce the redshifted $h\nu(1 - GM/c^2r)$.[21] The quantum nature of this energy ensures that this is equivalent to a time dilation which, to first order, is the same as the relativistic $\gamma = (1 - 2GM/c^2r)^{-1/2}$ and this, in general relativity, is the term responsible for the 'classical' half of the light bending.

If Newton did not produce a calculation of the gravitational bending of light, there exists in his manuscripts a calculation of atmospheric refraction which is similar in principle. Here Newton assumes a constant refracting field f at a height h above the Earth's surface, entirely analagous to the gravitational field g ($= GM/r^2$). He then uses the 'conservation of energy' equation, Proposition XLI, to calculate the resulting deflection into parabolic orbits of light rays entering the Earth's atmosphere.[22] The assumption of parabolic orbits requires mc^2 to be equated to the potential energy term $mfr\,(1 + \cos\phi)$, equivalent to the gravitational $GMm\,(1 + \cos\phi)/r$, while the use of Proposition XLI is equivalent to a modification of c^2 by the factor $(1 - 2fh/c^2)$ in the same way as the principle of equivalence is used to modify c^2 by $(1 - 2gr/c^2)$ or γ^{-2} in gravitational bending. In relativity this modification makes the term $c^2(dt)^2$ into $\gamma^{-2}c^2(dt)^2$ which, with c assumed constant, is interpreted as causing a dilation in measured values of time.

The force calculations of Cavendish and Soldner using hyperbolic orbits are therefore equivalent in principle to Newton's force calculations in refraction using parabolic or circular orbits and the potential energy terms which they imply for the processes are essentially the same. They do not use any principle involving a relativistic connection between space and time and it is *precisely for this reason* that they fail to predict the full bending of light. This also applies to calculations based on the principle of equivalence, because that principle effectively assumes infinitely fast particles and noncontracted elements of space.[23] This is illustrated by a comparison between the 4-vector line elements in the presence and absence of a gravitational field:

$$(ds)^2 = \gamma^2(dr)^2 + r^2(d\theta)^2 + r^2\sin^2\theta(d\phi)^2 - \gamma^{-2}c^2(dt)^2$$

and

$$(ds)^2 = (dr)^2 + r^2(d\theta)^2 + r^2\sin^2\theta\,(d\phi)^2 - c^2(dt)^2\quad .$$

Applying the factor γ^{-2} to $c^2(dt)^2$ without also applying that of γ^2 to $(dr)^2$ is a denial of the 4-vector connection between them.

It is clear from various attempts at deriving the full bending of light and the related effect of planetary perihelion precession, that the principle of equivalence, as a local rather than global phenomenon, is an inadequate heuristic medium for incorporating relativity or 4-vector theory into physics. Comparing classical and relativistic results we may say that the former give exactly what we would expect *by assuming $E = mc^2$ without incorporating the 4-vector mechanism.* Einstein's 1911 calculation turns out to be completely nonrelativistic!

In effect, then, the classical calculation of light bending shows that mc^2 is a potential energy term in classical physics which has the same effect as the equation $E = mc^2$ in relativistic physics, and that all the effects which depend only on $E = mc^2$ and not specifically on the 4-vector combination of space and time can be derived by classical approaches entirely independent of any concept of relativity. The true nature of the contributions made by different causes to the three relativistic predictions of redshift, light bending and perihelion precession have been obscured by the all-embracing nature of the general relativistic formalism, and it is too easily assumed that the effects can be derived only from the full field equations of general relativity. Comparison with classical predictions should now demonstrate that redshift and the time dilation components of light bending and perihelion precession depend only on the relation $E = mc^2$ and not on the 4-vector combination of space and time. The spatial components of the light bending and perihelion precession should then follow automatically from the application of 4-vector space-time without any need to apply the equivalence principle, any time dilation necessarily requiring an equivalent length contraction.

6. Newton's calculations on refraction

If Newton's expression for the force of refraction at least *implies* a potential energy term of the form mc^2 for light, it is his *quantitative* analysis of the phenomenon which brings him close to the modern understanding of the concept of mass-energy. Query 30 specifically invoked a *transformation* process between light and matter whose origin lay in Newton's realisation of the immense strength of the force involved in refraction, especially when compared with the force of gravity acting on the heavenly bodies. A force of such strength could, he believed, surely effect the transformation which his unitary theory of nature seemed to demand. He was well aware, as we are today, that one of the most important manifestations of this strength was the very great size of terms with c^2 as a factor, and if he did not refer directly to light particles as sources of 'energy', he did consider them as sources of momentum, a term which in relativistic quantum physics is entirely equivalent, and probably more significant.

The earliest calculation of the force mc^2/r involved in refraction appeared in a draft for Query 22 of the Latin *Opticks* of 1706 (Cambridge University Library, Add. MS 3970, f.292^v). In the same manuscript draft Newton also asserted that it was the small size of the particles of light which made them particularly active and, indeed, the chief source of activity within matter: "Now since light is the most active of all bodies known to us, & enters the composition of all natural bodies, why may it not be the chief principle of activity in them? Attraction in bodies of the same kind & vertue ought to be strongest in the smallest particles in proportion to their bulk. Tis much stronger in small magnets in proportion to their bulk then in great ones. For the parts of small ones being closer together unite their forces more easily."[24]

The version later published, which was transferred to the new Query 21 in 1717, further elaborated on this argument: "As Attraction is stronger in small Magnets than in great ones in proportion to their Bulk, and Gravity is greater in the Surfaces of small Planets than in those of great ones in proportion to their bulk, and small Bodies are agitated much more by electric attraction than great ones; so the smallness of the Rays of Light may contribute very much to the power of the Agent by which they are refracted."[25] And in a later draft of the refraction calculation (Add. MS 3970, f.621^{r}), he wrote: "And therefore since the rays of light are the smallest bodies known to us (For I do not here consider the particles of aether) we may expect to find their attraction very strong."[26]

Yet another argument, which found its way this time into Query 31, originated with this first calculation of refraction. Immediately below the calculation on f.292^{v}, Newton drafted a passage which in the printed version later became: "Now the smallest Particles of Matter may cohere by the strongest Attractions, and compose bigger Particles of weaker Virtue; and many of these may cohere and compose bigger Particles whose Virtue is still weaker, and so on for divers Successions, until the Progression end in the biggest Particles on which the Operations in Chymistry, and the Colours of natural Bodies depend, and which by cohering compose Bodies of a sensible Magnitude."[27] Z. Bechler has discussed how this hierarchy of particle structures and cohesive interactions could arise "through a kind of screening effect, in which the bulk of the particles could screen their neighbours' forces on account of their short range".[28] Newton had already considered such effects as early as the *Principia*,[29] and an early draft intended for a "fourth book" of the *Opticks* which was never published[30] indicates that he had calculated that the critical case involved forces inversely proportional to powers of the molecular separation greater than 4; very similar arguments are still used today. According to Bechler's analysis, this led to forces on composite bodies which were nonlinear functions of mass, explaining why the smallest bodies were subject to the strongest forces; and allowed Newton to propose a model of chromatic dispersion based on the mass differences between the particles responsible for light of different colours.[31]

The hierarchy proposal also recalls a passage in the suppressed *Conclusio* to the *Principia*: "By the ... motion of fermentation bodies can expel certain particles, which thereupon by their repulsive forces are caused to recede from each other violently; if they are denser they constitute vapours, exhalations and air; if on the other hand they are very small they are transformed into light."[32] We have already seen how

Newton effectively calculated the relative energies per unit mass associated with gas molecules and light particles in Query 21. It is therefore of especial interest that the first version of this argument, with the specific use of the relation

$$\text{velocity}^2 = \frac{\text{elasticity}}{\text{density}}$$

occurs immediately after the revised refraction calculation on f.622^v.[33]

The calculation of the type of force per unit mass involved in refraction suggested that it must be immensely more powerful than the force per unit mass due to the gravitation of the Earth or the force per unit mass due to the gravitation of the Sun. The calculation took some years to perfect and the early versions were vitiated by arithmetical and conceptual errors but eventually Newton was able to write in a draft of Query 22/30 that: "upon a fair computation it will (be) found that the gravity of our earth towards the Sun in proportion to the quantity of its matter is above ten hundred million of millions of millions of millions of times less then the force by wch a ray of light in entering into glass or crystal is drawn or impelled towards the refracting body. ... For the velocity of light is to the velocity of Earth in Orbis magnus as 58 days of time (in which) the Earth describes the (same space –); that is an arch equal to the radius of its orb to about 7 minutes, the time in wch light comes from (the Sun) to us; that is as about 12,000 to 1. And the radius of the curvity of a ray of light during it(s) refraction at the surface of glass on wch it falls very obliquely, is to the curvity of the earth Orb, as the radius of that Orb to the radius of curvature of the ray or as above 1,000,000,000,000,000,000 to 1. And the force wch bends the ray is to the force wch keeps the earth or any Projectile in its orb or line of Projection in a ratio compounded of the duplicate ratio of the velocities & the ratio of the curvities of the lines of projection."[34]

In another calculation in the same manuscript, Newton takes the radius of the Earth's orbit as 69 million miles (based on a solar parallax of 12 seconds) and the radius of curvature of the path of a light particle as 10^{-6} inches. Assuming that the light from the Sun takes 7.5 minutes to reach the Earth and that in this time the Earth would have travelled 6197 miles, he finds the ratio of the forces to be about 5×10^{26}.[35] Newton's ratio of 10^{26} or 10^{27} was an unprecedented figure for an order of magnitude calculation in physics, and became the first of the so-called 'large numbers' representing the ratios of events on astronomical and atomic scales.

7. Capillary action and the mass of a particle of light

Though Newton calculated the refractive force per unit mass over a radius of curvature of 10^{-6} inches, his statement that the force ratio was greater than 10^{27} implied that the actual radius of curvature was probably less than 5×10^{-7} inches and, therefore, more than an order of magnitude below the optical wavelengths (10^{-5} inches) which he had used in the early versions of the calculation (f.292^{v}) but of the same order as the molecular distances which he had found from experiments on pressing glass slides and lenses together. This suggests that, by the time of the revised calculation at least, he had come to believe that the force of refraction took place at the *molecular* level.

Now, subsequent to his calculation of the force per unit mass involved in refraction, Newton directed Hauksbee to perform an experiment on capillary action which enabled him to make a calculation of the absolute force of cohesion at molecular distances. Since he believed, at this time, that the forces involved in refraction and cohesion had essentially the same (electrical) origin, it is conceivable that he could have combined the two measurements to give a first order estimate of the mass of a particle of light. Simple calculations suggest that the value would have been something like 10^{-35} or 10^{-36} kg, which is close to the observed range of relativistic masses for photons in the visible region of the spectrum (0.32 to 0.55×10^{-35} kg); such a result would have been one of the most remarkable in the whole history of physics.

From Hauksbee's data in the capillary experiment, which had involved the measurement of the attraction between oil of oranges and glass, Newton calculated that: "the force of attraction is equal to the weight of a cylinder of the oil whose base is the same as the base of the drop and whose height is to 33.44/80 inches as 1/80 is to the thickness of the drop. Let the thickness of the drop be 10^{-7} inches and the height of the cylinder will be $33.40 \times 10^{5}/64$ or 52250 inches, which is 871 paces; and the weight of this will equal the force of attraction."[36] The thickness of 10^{-7} inches, though introduced into the account by way of illustration, was specifically chosen as representing the typical distance over which cohesive forces could be expected to act, and it is clear that Newton intended his calculation to represent the value of force at the *molecular* level. In presenting the experiment in Query 31 Newton first calculated the force at one inch diameter and then asserted that: "And where it is of a less thickness the attraction may be proportionately greater, and continue to increase, until the thickness do not exceed that of a single particle of the Oil."[37]

Also, since the intermolecular distance was derived directly from the study of optical interference phenomena, the force of capillary action at this distance could be compared directly with the optical force of refraction. The revised draft of the refraction calculation on f.621^r is immediately followed by a statement about the capillary forces involved in the Torricellian experiment which was surely intended to link the two forces on account of their short range and great strength: "The Atmosphere by its weight presses the Quicksilver into the glass, to the height of 29 or 30 Inches. And some other Agent raises it higher, not by pressing it into the glass, but by making its parts stick to the glass, & to one another."[38] In 1717 this was incorporated directly into Query 31 almost immediately preceding the description of Hauksbee's experiments.

Now, according to Newton's account, a drop of oil of diameter 10^{-7} inches will support a cylinder of volume

$$\frac{\pi \times 10^{-14} \times 52250}{4} \text{ cubic inches}$$

or 6.72×10^{-15} m^3. Taking the density of the oil as 842 kg m^{-3}, which is that of d-limonene, its major constituent, this is equivalent to a weight of 5.55×10^{-11} N. (From the density and chemical formula ($C_{10}H_{16}$) of d-limonene, we may estimate the separation of the individual oil molecules as 6.5×10^{-10} m, which is only four times smaller than the maximum of 10^{-7} inches which Newton had derived from the separation of contiguous glass slides.)

The inverse proportionality between force and distance for capillary action at small separations is equivalent to a constant value for force x distance, or energy of molecular interaction. Using Newton's values for force at 10^{-7} inches, this may be calculated at $5.55 \times 10^{-11} \times 2.54 \times 10^{-9} = 1.4 \times 10^{-19}$ J or 0.9 eV, which is reasonably close to electromagnetic energies in the visible region of the spectrum (1.8 to 3 eV). Newton's values are also equivalent to 2.8×10^{-2} Nm^{-1} for the surface tension of the oil/glass boundary. A constant value for the intermolecular potential energy in *gases* had been implied in Newton's earlier derivation of Boyle's law.

To compare the force of capillary action with the force per unit mass involved in refraction c^2/r, we need to take both forces over approximately the same distance r, and, from Newton's figures, we may suppose this to lie somewhere between 10^{-7} and 5×10^{-7} inches. For a distance r of 10^{-7} inches we obtain

$$c^2/r = 2.4 \times 10 \text{ Nkg}^{-1}$$

since the speed at which light travels 69 million miles or 1.11×10^{11} m in 7.5 minutes is 2.47×10^8 ms^{-1} Assuming, now, that essentially the same force is responsible for both capillary action and refraction, we obtain

$$\frac{5.55 \times 10^{-11}}{2.4 \times 10^{25}} = 0.23 \times 10^{-35} \text{ kg}$$

for the mass of a particle of light. For $r = 2 \times 10^{-7}$ inches the mass becomes 0.91×10^{-35} kg, and for $r = 3 \times 10^{-7}$ inches the mass is 2.1×10^{-35} kg.

Alternatively, we can make a more direct use of Newton's data and take the ratio of 10^{27} as a lower limit on the ratio between c^2/r and the force per unit mass exerted by the Sun on the Earth. For a time period T of 1 year (3.16×10^7 s) and solar distance R, this means that the force per unit mass involved in refraction is

$$\frac{4\pi^2 R}{T^2} \times 10^{27} = \frac{4\pi^2 \times 1.13 \times 10^{11} \times 10^{27}}{(3.16 \times 10^7)^2}$$

This gives us a lower limit of 4.39×10^{24} N kg^{-1} on c^2/r and an upper limit of 1.26×10^{-35} kg on the mass of a light particle.

8. The corpuscular and quantum theories of light

It is worth noticing, in relation to this reconstructed calculation of the mass of a light particle, that Newton's 'corpuscular' theory of light was much less mechanistic and much closer to the modern quantum theory than is commonly realised. This was a consequence of the fact that Newton's preferred method of reasoning outside the *Principia* was based on analytical abstraction from experimental results, rather than on synthetic development from a presumed mechanistic or 'hypothetical' model, however mathematically sophisticated this might be. "I said indeed", he wrote in an early letter to Oldenburg, "that the *Science of Colours was Mathematicall & as certain as any other part of Optiques*; but who knows not that Optiques & many other Mathematicall Sciences depend as well on Physicall Principles as on Mathematicall Demonstrations: And the absolute certainty of a Science cannot exceed the certainty of its Principles."[40]

He had observed periodicity and measured it using interference phenomena in thin films, and eventually convinced himself after further experiments (involving thick films) that periodicity was almost an *innate* property of light. "Light", he said, "is in Fits of easy Reflexion and easy Transmission, before its Incidence on transparent Bodies. And probably it is put into such fits at its first emission from luminous Bodies, and continues in them during all its progress."[41] However, his early optical work had already demonstrated that the property by which light produced the sensation of a particular colour was unchangeable and not subject to modification during reflection or refraction, and this, he was certain, must indicate a particulate nature for light. As he wrote on f.291[r], "Colours & refractions depend not on new modifications of light but on the original & unchangeable properties of its rays & such properties are best conserved in bodies projected."[42] Since periodicity was directly associated with the production of colour in Newton's experiments, this meant that some kind of periodicity or wave-like behaviour had to be associated with the actual particles of which light was composed. Newton did not eliminate this extraordinary ambiguity from his work precisely because *it was present in the data itself*; no other seventeenth century scientist showed such an uncompromising attitude to the primacy of experimental evidence.

Again, Newton's early experimental work in optics led him to regard the processes of reflection and refraction by mechanisms which in modern terms would be described as *field*-induced effects: "the Reflexion of a Ray", he wrote, "is effected, not by a single point of the reflecting Body, but by some power of the Body which is evenly diffused all over its Surface, and by which it acts upon the Ray without immediate Contact."[43] The same also applied to refraction, for "Bodies reflect and refract light by one and the same power, variously exercised in various Circumstances."[44] Originally, the reflecting and refracting power was conceived in terms of a mechanical aether,[45] but during the composition of the *Principia* (1684–7), the aether concept was abandoned in favour of forces acting at a distance. In reinterpreting his optical work of the 1666s and 1670s in manuscripts of the 1690s and 1700s, Newton worked by analogy with the gravitational force at the Earth's surface, which, as the collective effect of the Earth's individual particles, he was already explaining in terms of field theory.

It was the application of a field approach in Newton's refraction calculations which made this work directly comparable with classical energy calculations based on the summation of scalar potentials. In conjunction with the property of periodicity it led eventually to an understanding of the nonlinearities in the optical force which Newton observed but was unable fully to understand.

Newton's very ambiguity over the nature of light and the peculiar nature of the fits of easy reflection and easy transmission to which he supposed light rays were subjected took on new meaning when dualistic theories were finally forced upon a reluctant community of physicists. Newton anticipated the twentieth century, not because of some inexplicable 'coincidence' in viewpoint, but because he took almost the same positivistic attitude to the same experimental evidence; his ambiguity was the same as *our* ambiguity because it was based on experimental evidence which has remained unchanged to this day, and his presentation was almost devoid of theoretical content. Consequently, his light corpuscles, though possessing mass and momentum, were nothing like the particles of mechanistic philosophies and even their particulate nature was nothing but a necessary consequence of certain experimental results and not due to the assumption of any particular hypothetical model. The 'fits' have, indeed, been likened to the transition probabilities or probability wave amplitudes of quantum mechanics; and the expressions 'easy transmission' and 'easy reflection' seem to imply a probabilistic, rather than deterministic, interpretation. A memorandum by David Gregory of May 1694 records Newton's opinion that "A ray of light has paroxysms of reflexion or refraction, *and indeterminate ones at that*"[46] – a remarkable choice of words.

The explanation of interference rings in the early optical paper "Of ye coloured circles twixt two contiguous glasses"[47] is very much in the spirit of quantum theory with the momentum (p = mc) taken as the quantity determining the strength of the blow produced by the light corpuscle. The modern reasoning is that, for an air film of thickness t and light of wavelength λ at an angle r to the vertical, dark rings occur when

$$2t \cos r = m\lambda$$

and light rings when

$$2t \cos r = (m + 1/2)\lambda$$

where m represents the integer series 0,1,2,3.... Now, from quantum theory, the momentum of a light photon is given by

$$p = \frac{h}{\lambda}$$

where h is Planck's constant. From these equations we may derive the fact that the thickness of the film required to produce a given ring at any particular orientation of a light ray is inversely proportional to the momentum associated with the photons in the ray; and from the standard theory of the phenomenon the thickness of the film producing any ring is proportional to the area of the ring or its diameter squared. That is,

$$t \propto D^2 \propto \frac{1}{p \cos r} .$$

Newton was, therefore, completely correct, in modern terms, to propose that the size of the rings (and, by implication, the wavelengths) associated with light of different colours and incident at various angles varied inversely with the perpendicular components of the momentum of the relevant light particles, and his data showed a clear inverse proportionality between film thickness or area of ring and vertical component of momentum.

For some reason, he specified diameter, rather than area, for this proportionality relation in the case of obliquity. But he also considered the case for different diffracting media, and this time he actually specified the inverse proportionality of area or thickness (i.e. wavelength) and "motion" (i.e. momentum).[48] In this case the wavelength is reduced in the medium by the factor n, where n (the refractive index) is the ratio of the velocity of light in vacuum (c) to the wave velocity in the medium (v). As the wavelength is reduced, so the momentum, which is the ratio of energy to wave velocity, is increased. Assuming constant frequency ν,

$$p = \frac{h}{\lambda} = \frac{h\nu n}{c} = \frac{h\nu}{v}$$

and

$$E = pv \quad .$$

Newton writes: "if y^e^ medium twixt y^e^ glasses bee changed y^e^ bignes of y^e^ circles are also changed. Namely to an eye held perpendicularly over them, the difference of their areas (or y^e^ thicknesses of y^e^ interjected medium belonging to each circle) are reciprocally as y^e^ subtlity of ye interjected medium or as ye motions of ye rays in that medium."[49]

The material thus existed in Newton's early optical work for deriving the inverse proportionality between wavelength and momentum characteristic of the quantum theory; and it would be possible to derive a value of Planck's constant from Newton's data of the same order as the accepted value, say within the range 10^{-34} to 10^{-33} Js. For example, using 10^{-35} kg for mass and 2.47×10^8 ms for c, and calculating λ from the data in *Opticks* as lying between 2.0×10^{-7} m for violet and 3.2×10^{-7} m for red,[50] we obtain values of $mc\lambda$ (our h/2) between 4.9×10^{-34} and 7.9×10^{-34} Js. Unlike the calculation of the mass of a light particle, this could not really have been carried out by Newton himself, because the early work – which, for all its inconsistency of statement, did establish a connection between inverse wavelength and momentum – was set aside when he came to publish his *Opticks* and compose the Queries, apparently because it turned out to be incompatible with his deep-rooted belief that the largest light particles must also be the strongest.

The issue was complicated by the fact that light, as a wave motion, has *two* velocities, the velocity at which individual waves travel (the wave velocity) and the velocity at which the 'rays' travel (the group velocity). Newton's own view of light was similarly complicated, because he supposed his light vibrations to travel faster than the rays with which they were associated, and this may explain why he did not draw the obvious conclusion from his experiments that longer wavelengths implied faster rays. He was clearly aware of complications in the refraction process – ones due to the fact that momentum is not simply defined as the product of mass and velocity – and he eventually went so far as to speculate that the refractive force was a nonlinear function of mass, an idea associated in his manuscripts with the interconversion of light and matter.

Whatever differences may exist, however, there is at least one aspect of Newton's corpuscular theory which is fundamental to the way in which quantum theory is viewed today. This stems from his belief that the fixed and unchangeable nature of the rays of light, through all the processes to which they were subjected, must *necessarily* imply that they were composed of fixed and unchangeable particles of matter. Not only is this idea an essential component of modern theories of the interaction of light quanta and matter; it is also the basis on which the corpuscular theory and the quantum theory were used at different times to account for the same experimental evidence.

In 1832 David Brewster announced to the British Association his discovery that the absorption spectrum of nitrogen dioxide was crossed with more than a thousand identifiable dark lines. Needing to defend Newton against the proponents of the increasingly successful wave theory, he seized on this evidence to say that continuous waves could never explain a discontinuous spectrum. Line spectra, in fact, were never satisfactorily accommodated to continuous physics and this fact certainly formed part of the background to Bohr's brilliant application of the quantum theory of Planck and Einstein to Rutherford's model of the atom.

The great success of Bohr's theory was, of course, the explanation of the emission spectrum of hydrogen, and so, even if there had been no direct link between the old particle theory and the new, they were certainly both used to explain the same evidence – the line spectra of gases – and for essentially the same reason. In particular, the Bohr theory assumed that orbitals of electrons were fixed because the quanta of particular frequencies involved in electron transitions were fixed and could not be replaced by the equivalent amount of 'continuous' energy; Newton had been right, after all, in assuming that light was particulate because the rays of each kind had fixed and immutable properties.

9. Conclusion

It was Newton's unitary view of nature, with its strongly alchemical background, which made him insist that light was corpuscular despite his own evidence that it had periodic or wave-like properties. It was unlikely, he believed, that, if matter was made of corpuscles, light was not; there could not be two fundamentally different types of substance. He repeated the phrase "if nature be simple and conformable to herself" time after time throughout his writings and used it on numerous occasions to great effect. It is interesting to reflect, therefore, that there could, indeed, hardly be a more perfect illustration of Newton's principle of conformability in nature than de Broglie's original prediction of matter waves from which quantum mechanics subsequently developed. If waves of light have particle-like properties, and if nature is simple and conformable, then particles of matter should have wave-like properties.

It was the same unitary view of nature which made him prepared to consider even the interconversion of matter and light. Bodies certainly affected light, bending its rays, reflecting, refracting, diffracting, absorbing or emitting them; and light affected bodies, probably being "the chief principle of activity in them", since the absorption or emission of light was involved in so many other processes. The early versions of Query 22/30 stressed the strong connection between the transmutability or interconvertibility of light and matter and the role of light as the 'active power' or, as we should say, 'energy source' within matter: "Do not bodies & light mutually change into one another. And may not bodies receive their most active powers from the particles of light wch enter their composition (of wch they are composed *cancelled*)?"[51] A unitary view also required that interconversion was a *mutual* process – particles of light not only possessed mass but actually made up matter – and it is from precisely the same reasons that we now infer that the equation $E = mc^2$ applies not only to light but also to 'solid' matter.

However, it was a series of precise calculations, rather than merely qualitative arguments, which made Newton believe that the electro-optic force was a possible mechanism for the transition. The appearance of ratios and terms involving c^2 or mc^2 in these calculations should not be seen as bearing only a coincidental resemblance to modern ideas. In principle the concepts used by Newton and by modern physicists are exactly the same, even though the latter are not normally aware of it. If Newton tends to use terminology which is unfamiliar to us and to place emphasis on different aspects of similar material – principally because he concentrated on force rather than energy – his fundamental line of reasoning is in no way different to that of the modern theorists; and the current preference for momentum, rather than energy, as the basic concept in relativistic quantum theory provides a strong point of contact. It was one of Newton's greatest strengths that he was able to develop physical theories that were purely abstract and so applicable to areas well beyond the ones with which he was immediately concerned; it has been a weakness of modern physics that it has not always been prepared to recognise its own abstract basis.

It is no coincidence, therefore, that the draft Query in which Newton produced the calculation of the force per unit mass ratio for the refraction of light was also the one in which he first argued that light could be converted into matter, and matter into light, and further declared that light was the active principle in all organised matter.[52] His calculation of the very great strength of the refracting force was specifically intended to show that it was strong enough to effect the transmutation. According to Bechler, the whole argument led to the statement, now in Query 31, but originally part of this draft, that matter was organised at a succession of different levels, with weaker forces at successively higher levels of organisation; and it stemmed from a mass model of dispersion in which the largest particles were the least accelerated,[53] an idea which may be taken as a partial return to his earlier views on the inverse relationship between wavelength and momentum.

Because he split up this draft between several Queries and suppressed some of the material altogether, Newton gave the impression that its various conclusions were merely oracular pronouncements on various subjects of interest, prescient to be sure, but not stemming from any unified theory of light and matter, and bearing only a coincidental resemblance to certain modern developments.

Far from being merely brilliant speculative guesses, however, Newton's views on the relation between light and matter, and between force and material structure, are exact consequences of his corpuscular theory of light, backed up with precise mathematical calculations which differ in no significant respect from those based on special relativity and quantum theory. Despite superficial differences in their presentation as mechanistic models, the Newtonian and relativistic quantum theories of the mass-energy of the photon yield exactly the same values, in principle, for significant quantities because they are both based on the same more fundamental abstract mathematical truths.

Notes and references

1 I. Newton, *Opticks* (preface by I. B. Cohen, foreword by A. Einstein, introduction by E. T. Whittaker), Dover Publications, 1952 (based on the fourth edition, 1730), 374

2 R. S. Westfall, *Force in Newton's Physics*, Macdonald, 1971, 487

3 E. J. Aiton, 'The Inverse Problem of Central Forces', *Annals of Science*, 20, 81-99, 1964, 82

4 Prop. XLI is discussed in D. T. Whiteside, 'The Mathematical Principles Underlying Newton's *Principia*', *Journal for the History of Astronomy*, 1, 116-38, 1970

5 Westfall, op. cit., 1971, 490

6 H. W. Turnbull, J. P. Scott, A. R. Hall and L. Tilling, *The Correspondence of Isaac Newton*, 7 vols., Cambridge University Press, 1959-1977, 3:384

7 H. G. Alexander, *The Leibniz-Clarke Correspondence*, Manchester University Press, 1956, xxxf, 121-5

8 Cambridge University Library Add. MS 3975, p.10, section 34, quoted by R. S. Westfall, "Isaac Newton's Coloured Circles twixt two Contiguous Glasses", *Archive for History of Exact Sciences*, 2, 181-196, 1965, 184

9 J. Stachel and R. Toretti, 'Einstein's first derivation of mass-energy equivalence', *American J. Phys.*, 50, 760-3, 1982, 761

10 A. Einstein, *Jahrb. Radioakt.*, 4, 440ff, 1907, 442; quoted Stachel and Toretti, op. cit., 761

11 Newton, *Opticks*, op. cit., 351

12 The very close numerical agreement in the values resulting from this essentially approximate calculation stems from the fact that Newton's error in making the velocity squared ratio of light to sound too low by a factor of 1.5 is compensated for by the factor of 1.4 which is introduced by taking the velocity of sound as the molecular velocity.

13 *Principia*, Book II, Prop. 23, Theor. 18; I. Newton, *Mathematical Principles of Natural Philosophy*, translated by A. Motte, revised by F. Cajori, Cambridge University Press, 1934, 302

14 J. Herivel, 'Newton and the Law of Centrifugal Force', *Isis*, 51, 546-52, 1960; the connection with the kinetic theory of gases is mentioned by S. G. Brush in *The Kind of Motion We Call Heat*, Elsevier, 1976, 15

15 L. Rosenfeld, 'Newton and the Law of Gravitation', *Archive for History of Exact Sciences*, 2, 365-86, 1962-5, 373

16 A. R. Hall and M. B. Hall, *Unpublished Scientific Papers of Isaac Newton*, Cambridge University Press, 1962, 312-7

17 Of course, the 'coincidence' is really an expression of the fact that inverse-square laws are a natural result of three-dimensional space, inverse-square-law forces between elements leading to constant force on integration.

18 Its mathematical equivalence to yet another expression, that for the velocity of waves on a vibrating string, seems to have been responsible for Newton's extraordinary assertion in the famous 'classical scholia' that the ancients had discovered the law of gravity. See J. E. McGuire and P. M. Rattansi, "Newton and the 'Pipes of Pan'", *Notes and Records of the Royal Society*, 21, 108-143, 1966.

19 C. M. Will, 'Henry Cavendish, Johann von Soldner, and the deflection of light', *Amer. J. Phys.*, 56, 413-5, 1988

20 A. Einstein, *Annalen der Phys.*, 35, 898, 1911
21 Goldstein's more accurate analysis, giving second order terms, takes hv as a function of R, but second order effects (at microseconds of arc in the case of deflection) are well beyond the limits of current detectability. (J. Goldstein, 'A close look at a canonical semiclassical derivation of the gravitational redshift', *Amer. J. Phys.*, 55, 476-7, 1987)
22 See D. T. Whiteside, *The Mathematical Papers of Isaac Newton*, 8 vols., Cambridge University Press, 6:344-56, and 'Kepler, Newton and Flamsteed on Refraction Through a "Regular Aire": the Mathematical and the Practical', *Centaurus*, 24, 288-315, 1980; and Z. Bechler, "Newton's Search for a Mechanistic Model of Colour Dispersion: A Suggested Interpretation", *Archive for History of Exact Sciences*, 11, 1-37, 1973, 24-5.
23 See, for example, R. P. Cromer and J. D. Lathrop, 'The principle of equivalence and the deflection of light by the Sun', *Amer. J. Phys.*, 46, 801, 1978.
24 Quoted Z. Bechler, "Newton's law of forces which are inversely as the mass: a suggested interpretation of his later efforts to normalise a mechanistic model of optical dispersion", *Centaurus*, 18, 184-222, 1974, 201-2
25 Newton, *Opticks*, op. cit., 351-2
26 Quoted Bechler, 1974, op. cit., 207
27 Newton, *Opticks*, op. cit., 394
28 Bechler, 1974, op. cit., 198
29 *Principia*, Book III, Prop. 6, Coroll. 5, op. cit., 414; see also Book I, Props. 85-6, Theors. 42-3, op. cit., 214.
30 Cambridge University Library Add. MS 3970, f.336^{r}; I. B. Cohen, "Hypotheses in Newton's Philosophy", *Physis*, 8, 163-4, 1966, 180
31 Bechler, 1974, op. cit., 192ff
32 Hall and Hall, op. cit., 341-2
33 Bechler, 1974, op. cit., 221
34 Cambridge University Library Add. MS 3970, f.621^{r}; quoted Bechler, 1974, op. cit., 208
35 Cambridge University Library Add. MS 3970, f.621^{r}; Bechler, 1974, op. cit., 220
36 *De vi electrica*, c.1712; *Correspondence*, op. cit., 5:368
37 Newton, *Opticks*, op. cit., 393-4
38 Cambridge University Library Add. MS 3970, f.621^{r}; reproduced in Bechler, 1974, op. cit., 186. (Experiments by Hauksbee and others showed that shaking the mercury in a Torricellian tube caused emission of light at the same time as overcoming the force of cohesion, suggesting a possible link between optical and capillary effects.)
39 Using the value 6.5×10^{-10} m for the molecular diameter makes the potential energy for a single molecule of d-limonene something closer to 0.06 eV but we are assuming that Newton would have equated summations of the optical and capillary forces over similar distances. (According to further calculations on f.622, mentioned in Westfall, 1971, op. cit., 423, the equivalent molecular energy for water at 10^{-7} inches would be greater than 1.3 eV, and so even closer to the optical energy range.)
40 11 June 1672; *Correspondence*, op. cit., 1:368
41 Newton, *Opticks*, op. cit., 282
42 Newton, *Opticks*, op. cit., 266-7
43 Quoted Bechler, 1973, op. cit., 33

44 Newton, *Opticks*, op. cit., 269
45 For example, in 'An Hypothesis Explaining the Properties of Light'; *Correspondence*, op. cit., 1:362-85
46 D. Gregory, 'Annotations, Physical, Mathematical and Theological from Newton', 5, 6 and 7 May 1694; *Correspondence*, op. cit., 3:339
47 Cambridge University Library Add. MS 3970, ff.350-53^{v}, Westfall, 1965, op. cit., 191-6
48 Prop. 4, Westfall, 1965, op. cit., 191. It is worth noting here that, in this very early optical work, Newton *did* actually employ something resembling a conventional wave interpretation, based on the pulse theory of Robert Hooke.
49 Westfall, 1965, op. cit., 191
50 We may use the ratios given in *Opticks*, Book II, Part IV, Observation 8 to determine the 'intervals of fits' for red and violet rays from the value of 1/89 000 inch for those at the border of yellow and orange given in Book II, Part I, Obsevation 6. As Newton had no knowledge of the principle of interference, these 'intervals of fits' are only half the size of our wavelengths.
51 Cambridge University Library Add. MS 3970, f.292r; Bechler, 1974, op. cit., 200
52 His alchemical studies, reflected in the passage from the *Conclusio* quoted above, and also in such manuscripts as 'Of natures obvious laws & processes in vegetation' (Burndy MS 16), were, of course, already leading him in this direction.
53 Bechler, 1974, op. cit., 192ff

APPENDICES

1. Newton's analogy between optics and steady state dynamics

Formulae used by Newton (or equivalent)

Dynamics	Optics
orbital motion (Prop. XLI)	*atmospheric refraction*
$V = mv^2$	$V = mc^2$
orbital motion (gravity)	*refraction at surface*
$F = \frac{mv^2}{r}$	$F = \frac{mc^2}{r}$
molecular motion	*motion of light particles*
elasticity of aether $= \rho v^2$	elasticity of aether $= \rho c^2$
velocity of sound	*velocity of light*
$v = (k/\rho)^{1/2}$	$c = (k/\rho)^{1/2}$

Formulae used by Newton's successors

Dynamics	Optics
gases in 3-D	*radiation gas in 3-D*
$P = \frac{\rho v^2}{3}$	$P = \frac{\rho c^2}{3}$
virial theorem	*relativity*
$V = mv^2$	$E = mc^2$

2. Post-Newtonian derivations of light bending

classical	using principle of equivalence
optical PE = mc^2	$E = mc^2$
subtract gravitational PE	subtract gravitational PE
	gravitational redshift = time dilation
gravitational bending	gravitational bending

3. Newton's optical ideas: some quantitative and qualitative developments

Newton's rings experiment

estimation of molecular separation
$< 2.5 \times 10^{-9}$ m

force of capillary action over molecular diameter
5.6×10^{-11} N

capillary action and refraction caused by same (electrical) force at molecular level

mass of light particle
0.2 to 1.3×10^{-35} kg
or momentum
0.5 to 3×10^{-27} Ns

'intervals of fits' ($\lambda/2$)
2 to 3.2×10^{-7} m

refracting force per unit mass (c^2/r)
4×10^{24} to 2×10^{25} Nkg^{-1}

early statement: strength of ray measured as momentum x velocity
($pc = mc^2$)

early statement: momentum inversely proportional to interval of fits

(these ideas not pursued by Newton in later work)

4. The qualitative development of Newton's ideas on interconversion

assumed constant
velocity of light

analogy between optics
and steady state dynamics

refraction calculation

immense strength of
optical force
$F = mc^2/r$

hierarchy of particle
forces

formula for velocity of
waves
(light / sound)

elasticity per unit mass
of light / sound
(our energy per unit
mass ratio
4.9×10^{11})

hierarchy of organisation
in matter

light particles smallest
and most 'active' components
of matter

light particles
'make up' matter

light interconvertible
with matter